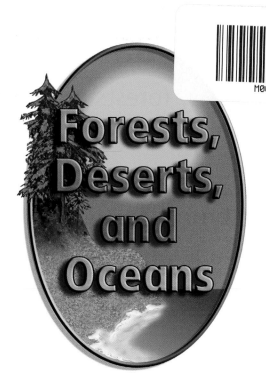

Forests, Deserts, and Oceans

Printed in Mexico

ISBN 978-0-15-362185-7

ISBN 0-15-362185-0

8 9 10   0908   16 15 14 13 12

4500375988

Harcourt
SCHOOL PUBLISHERS

Visit *The Learning Site!*
www.harcourtschool.com

The Forest Habitat

There are many forests around the world. A **forest** is land that is covered with trees.

Trees have branches and leaves. Trees make shade on the forest floor. Shade helps keep the soil moist.

This forest has many trees.

These ferns need water. They do not need much light.

Forest trees need rain to grow tall. Their leaves take in light. Light helps them grow, too.

Ferns and flowers grow under the trees. These smaller plants need water, but not as much light as trees.

 MAIN IDEA AND DETAILS Describe the forest floor.

Forest Animals

Many different animals live in the forest. The forest is their habitat. A **habitat** is a place where an animal finds food, water, and shelter.

Fast Fact

Young bears are called cubs.

Bears find food, water, and a place to sleep in this forest.

The legs on this beetle help it to burrow in wood to find food.

Bears need a large part of the forest for their habitat. They look for food and water in many different places in the forest.

Wood beetles do not need much more than a log for their habitat!

 MAIN IDEA AND DETAILS **What is your habitat? Tell where you find food, water, and shelter.**

The Desert Habitat

A **desert** is land that gets very little rain. The soil is very dry. It is sunny all year long. Few plants and animals can live in this habitat.

The Sonoran Desert is in Southern Arizona.

The desert has little water. This cactus uses its waxy covering to keep water in.

prickly pear cactus, Sonoran Desert

Living things can only survive in a place that gives them what they need to live. Desert plants do not need much water. They can live in a dry place such as the desert.

MAIN IDEA AND DETAILS Some plants need a lot of water. Why can the desert not be their habitat?

Desert Animals

Desert animals are special. They find ways to stay cool. They find food and water where it is very dry.

Coyotes can find water by digging. They often hunt at night when it is cool.

Fast Fact

Coyote melons are plants that hold water. They taste bad to most animals. Coyotes love them!

Coyotes have a strong sense of smell. This helps them find food.

This snake is cooling off under the sand.

Desert tortoises dig homes in the ground. This keeps them cool. They eat plants for their food. These same plants give them the water they need!

 MAIN IDEA AND DETAILS **Why does a desert tortoise need feet that can dig?**

The Ocean Habitat

An **ocean** is a large body of salt water. It is a habitat for fish, sea turtles, and other ocean animals. These animals find what they need in or near the salt water.

Fast Fact

A baleen whale eats 8,000 pounds (3,629 kilograms) of food a day. It has no teeth.

 MAIN IDEA AND DETAILS Where does a crab find its food?

The crab's legs help it get food.

This whale finds food in deep, cold water. A layer of fat keeps it warm.

Summary

A habitat is an animal's home. Animals live in the habitat where they can find food, water, and shelter. Some animals live in the forest. Some live in the desert. Some live in the ocean.

Glossary

desert A land that gets very little rain
(6, 7, 8, 9, 11)

forest A land that is covered with trees
(2, 3, 4, 5, 11)

habitat The place where an animal finds food, water, and shelter
(2, 4, 5, 6, 7, 10, 11)

ocean A large body of salt water
(10, 11)

*Freedom
from Want*

Children safe in bed showed *Freedom from Fear*. The last picture showed people of different faiths together. It was called *Freedom of Religion*.

These were some of the freedoms America was fighting to keep. Most of Rockwell's paintings were fun, but these paintings were very serious. He felt they showed the very best things about America.

Homecoming Marine cover for
The Saturday Evening Post, 1945

When the war was over, Rockwell painted many pictures of soldiers coming home. The soldiers had been gone a long time. Now they were war heroes.

Norman Rockwell and his wife, Mary, had three sons. He often asked his sons to be his models. He liked to draw using real people.

He paid children to pose for him. It was sometimes hard for them to stay still while he made his drawings. Rockwell also took photographs of people and painted from them.

Norman Rockwell's
paintings told stories of
everyday life and everyday
people. He also liked to
paint funny moments
and heroes and times
of adventure.

People liked the way
the pictures made them
feel. They thought Norman
Rockwell understood their
lives or the lives they
wished to have.

Rockwell wanted to get
every detail right in his
paintings. The people were
the most important part.
Their faces told the story
Rockwell wanted to paint.
Sometimes he would want
them to look happy, afraid,
or silly. He would show his
models the faces he wanted
them to make!

The years went on and times changed. Life
in America kept changing too. And Norman
Rockwell never stopped drawing and painting
what he saw around him. People were going to
outer space! Now he painted modern-day
heroes such as astronauts.

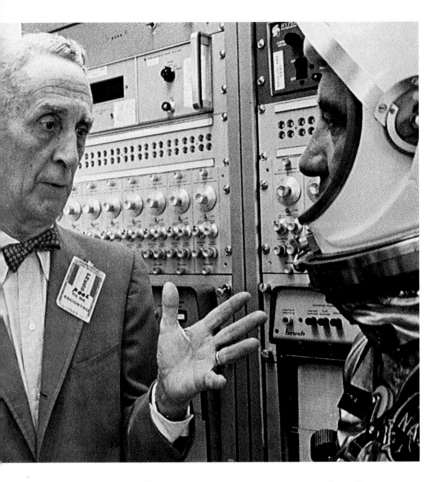

In 1969, the space program put the first astronaut on the moon. Rockwell knew there were many people who helped the astronauts get to the moon. When Rockwell painted this subject, he wanted everyone in it. He knew a whole team of scientists and others had made it happen. That's the story he wanted to tell.

Rockwell's art was admired and loved all over America. In 1977, President Ford gave Rockwell the President's Medal of Honor. By the time he died in 1978, Rockwell had painted more than four thousand works of art. Incredible!